"COUNT ON US"

COAST GUARD CUTTER DEPENDABLE

LAW ENFORCEMENT

AND

SEARCH & RESCUE

By

Edward Leo Semler Jr.

First Edition: 2021

Library of Congress Control Number: 2021915284

ISBN: 978-1-7376472-0-1

Printed in the United States of America

City of Publication: Schulenburg, Texas

Cover picture and layout by Edward Leo Semler Jr.

To everyone who has served on *Dependable*

Especially the 2003 to 2005 crew

TABLE OF CONTENTS

Introduction

This book is about my time aboard the *Dependable* from the 5th of June 2003 to the 1st of June 2005. At the time she was homeported out of Cape May, New Jersey along with her sister 210 the *Vigorous*. By the time I did my tour on *Dependable* I already had over 20 years in the military, and I guess in some ways I thought my career was winding down and in other ways it was just getting started. This was due to the fact that I was over the longevity time requirement to retire, but I had also just accepted a commission from an enlisted man to an officer. The latter required me to pay back that gift from the Coast Guard with a few more years of service.

Before I jump into this book it's probably a good time to just skim over the Coast Guards structure. It is a military organization and was under the Department of Transportation until just before I stepped aboard *Dependable*, changing to the Department of Homeland Security on the 1st of March 2003. Its rank, pay, and retirement structures are the same as the other Department of Defense services. The enlisted ranks run from E-1 to E-9 and the officer ranks from O-1 to O-10. With warrant officers ranging from W-1 to W-4. And retirement is generally obtained after 20 years of service.

My rating, or job specialty as an enlisted man was that of a machinery technician. It's sort of a jack of all trades in the mechanical engineering field and covers engines, gas turbines, basic electricity, refrigeration and air conditioning, hydraulics, and so on. I had done well over my 20 years and had obtained the rank of master chief, or E-9. But after the events of September 11th 2001 the Coast Guard needed to grow its officer core and solicited their enlisted ranks for applicants. I had my Bachelor's degree and decided to apply. I was accepted and commissioned as a lieutenant junior grade, O-2, in the engineering field. With my commission came orders to the *USCGC Dependable (WMEC-626)* as an engineering officer in training. This was to prepare me for a follow on tour as an engineering officer on another cutter.

But *Dependable* had way more service time than I had racked up and was already 35 years old when I walked across her gangway in 2003, which is considered ancient for a military service vessel. *Dependable* was the 12th cutter commissioned and built by the American Ship Building Company in Lorain, Ohio in late 1968. Her Reliance Class of 210 foot cutters, which totaled 16, were laid down by various shipyards to include the Coast Guard Yard, American Ship Building Company, Todd Shipyard, and the Christy Corporation. As I write this book in late 2021 she along with most of her sister ships are still active and patrolling the seas. Most are still with the United States Coast Guard and a few given to other countries. But their days of U.S. service are coming to an end.

As the Coast Guard builds their newer class of cutter they are decommissioning the 210's and passing them on to other country's to use in their Coast Guards and Navy's. So their service will continue. Just under a different name for a different country.

The class had gone through a major overhaul in the late 1980's and early 90's which standardized the main propulsion to 2 Alco 251 diesel engines, replaced the forward gun, and most significantly replaced the stern tube exhaust with exhaust stacks amidships. Their current configuration was with a helipad for helicopter operations, a port and starboard small boat, a forward mounted 25mm MK38 cannon, and two .50 caliber machine guns. Their relatively small crew of 75, compact size, max speed of 18 knots, and range of up to 9,000 nautical miles made them a very efficient asset.

***Dependable* in Boston, MA. when I reported aboard**

I arrived aboard *Dependable* on the 5th of June 2003. It was an easy transfer because I was currently station at the Naval Engineering Support Unit in Boston and just had to walk down the pier and hop aboard. She was conducting a normal two month D1 patrol in the Atlantic Ocean and had pulled into Boston on a port call. We referred to our patrols by what Coast Guard District we were operating in. A D1 patrol was the North Atlantic area of District 1, D5 was the Mid Atlantic, D7 the Southern part of the Atlantic and the Caribbean Sea, and D8 the Gulf of Mexico.

The *Dependable* had a crew of 75, 12 officers and 63 enlisted personnel. It was an all-male crew with the exception of two female junior officers. Because every person has a vital role to play on a cutter, berthing areas have to be kept at capacity. So if female officers were assigned they would be assigned in pairs. And at the time the majority of cutters, except for the larger 270' and above, could not handle female enlisted due to the lack of berthing space, or rather the lack of females to fill them.

And I know this might not be the most gender sensitive thing to say, although I did see in a recent Coast Guard publication an article praising an all-female aircraft crew, but I enjoyed an all-male crew. I had sailed on numerous all male cutters and mixed crew cutters. Personally, I felt a different interaction with these crews. And this is just my personal opinion but all male crews bonded together better and there was a more relaxed atmosphere. With mixed crews I felt that guy's personalities changed to impress the women and that you had to watch what was said as not to offend the opposite sex. And of course there were the courtships that inevitably went on.

Even on *Dependable* with just the two female junior officers it was noticeable. They were professional. But they were the same young age as most of the enlisted crew and when swim call was held the crew definitely noticed them in their two piece bikinis.

I can vividly remember on *Seneca*, a 100% all male crew, we had some female cadet's sail with us for a bit and were going to be standing some watches in the engine room. The first day they were on board I walked into the engine room control booth and a wave of aftershave greeted me. I had never smelled aftershave underway and was wondering what the heck was going on until I rounded the corner to find the aged electricians mate chief wearing the cleanest set of coveralls I ever saw him squeeze into, hair combed, and smelling like he got dipped into a vat of aftershave. And of course, I could write another book on the courtship issues I dealt with on the mixed crew of *Sherman*. And I by no means want to give the impression that females are any less a sailor. They are fantastic. But when you throw them together with males, things change. It's a different atmosphere and I preferred an all-male crew.

As a junior officer I was assigned to a berthing space forward and below the main deck. I shared it with another officer and the room had a bunk bed, two work station desks – one for each of us, its own shower, toilet, and sink. The berthing accommodations were some of the best aboard. Only the officers – the captain, executive officer, engineering officer, operations officer, and junior officers had their own, or two man, staterooms.

The eight chiefs, E-7's, slept in two separate four man staterooms with two bunk beds in each stateroom and shared a head with two sinks, toilets, and showers. The enlisted, E-1 thru E-6, were grouped into 10-12 man berthing areas with racks three high with attached heads containing several sinks, toilets and showers.

When we were inport the single crewmembers were given rooms in the barracks over on the base, which was a nice luxury. Back in my younger days, or what the old chiefs would refer to as "back in the old guard," the single crew lived on board the cutter all the time. But work-life studies rightfully showed that it was a strain and there was never time away from work and the cutter. Personally, I enjoyed it and lived on board all of the cutters I was stationed on except one, *Seneca*. Even here on *Dependable* I lived aboard the cutter. I would just go home on the weekends when I didn't have duty. I had always chosen to keep my family in the Washington, DC. & Baltimore metro area. It was easier on them because they weren't moving around every few years. And it was easier on me because I could concentrate on my duties and not be worried about getting home every night while in port. It also let me bond better with my co-workers and crew. This was actually a common thing to do and they called it being a "geographical bachelor."

The phrase "back in the old guard" was a phrase that use to befuddle me in my early Coast Guard years. The chiefs and senior enlisted were always throwing that phrase around when they wanted to sound salty and experienced. It's kind of like the stories you heard from your parents about them walking barefoot in the snow, up a hill both ways, to school. They always had it worse than you!

The Commanding Officer, or CO, was Commander, O-5, Mike Christian and he ran a fast paced operating tempo. I had previously served on five other cutters but had never conducted more boardings, migrant interdictions, and drug busts then on *Dependable*. As you read the patrol summaries which follow you'll see this cutter meant business! And this was only a 210 foot medium endurance cutter, the smallest law enforcement cutter I had served on.

Even though I was a junior officer Commander Christian respected me for the high enlisted rank I once held, something he didn't have to do. Over my two year tour he would interact with me more as an experienced senior enlisted man rather than a junior officer and I really appreciated that. He noticed that the shift from senior enlisted to junior officer was tough for me. He was a true mentor and his guidance helped me get through this first tour underway as an officer.

My primary duty was as an engineering officer in training, EOIT, but I would assume the duties of the damage control assistant, DCA, and electrical and damage control division officer. As I finished my qualifications I would assume other collateral duties such as the civil rights officer and voting officer. It was a small crew and everyone had to take on extra duties.

One of the other collateral duties I would be assigned from time to time was that of the investigating officer for individuals brought up on charges under the Uniform Code of Military Justice, UCMJ. It's the universal military legal code in the

U.S. Armed Forces. And I'm going to share a few of the cases I had to attend to so you can get a feel for discipline in the Coast Guard.

The process would start with an individual having a form CG4910 written up on them formally charging them with an offense. The slang term for this was "booking" someone. The executive officer, XO, would receive the CG4910 and assign an investigating officer to ascertain the facts and to see if the charges and facts met the requirements for an offense per the UCMJ manual. If they did, the matter was taken to captain's mast, the hearing in front of the captain. If not, the matter was dropped or handled with a verbal or written reprimand.

When going to Captains Mast it was held in a formal setting with the captain sitting behind a table covered with a green tablecloth. This green tablecloth was the time honored nautical tradition indicating that you were standing before the captain to answer for an offense. The accused would enter in dress uniform and stand in front of the sitting captain and hear the charge against them read. Usually, the accused brought in someone of prominence aboard the cutter to speak on their behalf, sort of like a character witness or lawyer.

The captain could also hold the mast in front of the crew if he wished. I have seen it done both ways. Usually this only happened when the captain wanted to press home a regulation or set an example. That way everyone saw what happened at the mast, the captain could make a statement in his sentencing, and there is no scuttlebutt as to what "happened in the mast."

Getting the Green tablecloth out for a Captains Mast

There were two incidents that were sort of typical of the charges brought up to mast here on *Dependable* - drugs and alcohol. I was assigned the perfecta of what started out as an alcohol incident and turned into an alcohol and drug incident.

A young seaman, E-3, from *Dependable* was driving through the front gate at Cape May Training Center, where *Dependable* was moored. He was maneuvering his vehicle around the jersey barriers leading up to the gate when he ran into one. The civilian officer at the gate walked up to the seaman's vehicle to investigate and found the seaman to be drunk. He was taken to the police station on base and given a breath test which he failed. He was brought up on charges, which were forwarded to the *Dependable*, and his car was impounded.

The XO assigned me as the investigating officer. The case seemed pretty cut and dry. I called up to the warrant officer in charge of the base police detachment and he said he would compile the paper work on the incident involving the seaman for me since he had to do a report anyway. He called me the next day and asked me if I wouldn't mind coming up to his office, he had something to show me. I arrived at the police station and the warrant officer handed me his file folder and a stack of pictures. He said, "We arrested him for DUI, but you may be interested in these pictures we found in the vehicle's glove box." As I shuffled through the pictures they were all of the seaman and other people smoking pot, or some sort of drug, out of a makeshift pipe. Needless to say the case went to mast and the seaman was found guilty of drug use and discharged from the service.

It got me thinking back to a time about 19 or so years earlier. On this same base I was caught on a urinalysis drug test and charged with the use of marijuana. My punishment then was working on base until I cleared my drug tests. After that I was sent on to my first duty assignment, the *Sumac*. Oh how the times had changed. I appreciated that the Coast Guard had taken the time to invest in me and allowed me a second chance. I would like to think they got a good return on their investment.

I was later assigned the case of another seaman involved in an alcohol incident. This one demonstrates the double jeopardy members face in the military judicial system that is not allowed in the civilian system.

In this case, a seaman was on leave driving a vehicle with several of his civilian friends as passengers. The group drove by a policeman who had another vehicle pulled over. As they drove by, one of the seaman's friends yelled out the window something to the effect of, "Fuck you pig!" The police officer got into his cruiser and pursued the seaman's vehicle, pulling him over. The seaman was given a breathalyzer test, which he failed. He was arrested and once the police discovered the seaman was in the service our command was notified. I was assigned as the investigating officer and I called the police officer who arrested the seaman. He informed me that the local police were going to press charges. I relayed this to the command, hoping we would drop our investigation. The command directed me to file a CG4910 on him and if my investigation turned up enough evidence to charge him, we would take him to mast.

The seaman was found guilty in the civilian court and had to pay fines. I found his actions in violation of the UCMJ and he was taken to mast where he was found guilty, received 30 days confinement, 30 days extra duty, and 30 days loss of pay from the Coast Guard! The 5th Amendment to the United States Constitution, which addresses being punished twice for the same crime, only applies to judicial criminal proceedings. The military loop hole is that all the punishment handed down at captains mast is deemed non-judicial under the UCMJ. This punishment can be handed down legally in conjunction with a civil conviction or a military judicial court martial.

Another interesting incident worth mentioning occurred when I was making rounds of the vessel with Captain Christian while underway. One evening, after the nightly department head report was given to the captain at 1900, he asked me to accompany him on a round of the cutter. It wasn't typical for the captain to want to tour the cutter after evening reports so I wasn't expecting it. We were underway so all the hatches leading below the main deck were secured and we had to open the small scuttles to transit into the lower spaces, which is very time consuming. Everything was going fine until we started to head toward engineering berthing, and I got a sudden feeling of dread in the pit of my stomach.

I just remembered that the engineers were planning a sombrero party for one of the guys leaving and I knew it was against regulations. I know I shouldn't have been turning a blind eye to the breaking of rules, but the crew had to have ways of blowing off steam while underway and this was truly a harmless tradition. The Coast Guard at the time was going through a hazing witch hunt and just slapping someone on the back to say welcome aboard was deemed a hazing infraction!

A sombrero party was a simple going away event in which a member transferring off the cutter was tied down, covered with shaving cream, and had a sombrero placed on his head. A picture was always taken and given to the departing member as a souvenir from his shipmates.

As the captain spun the handle on the scuttle to enter the engineering berthing space below, I was getting nervous. As soon as the captain climbed down through the small opening I hurried down the ladder behind him and shouted attention on deck to try and warn the unaware engineers of the pending danger.

It didn't work, and the captain came face to face with an engineer lying on the deck, wire tied to a table, covered in shaving cream, and wearing a sombrero!

If I were to say that the captain was pissed that would be an understatement! He wanted heads to roll and the usual engineering suspects, consisting of two of my electrician mates, EM3 John "Hank" Dudley and EM3 Matt "Juice" Lesniak, were brought up on hazing charges and taken to mast. They asked me to act as their character witness at mast and the investigating officer duties were handed off to another junior officer. With a lot of fast talking, and pleading for compassion, I managed to get their charges reduced and they were placed on probation. I also appreciated that they didn't throw me under the bus in front of the captain and said "but Mr. Semler knew about it!"

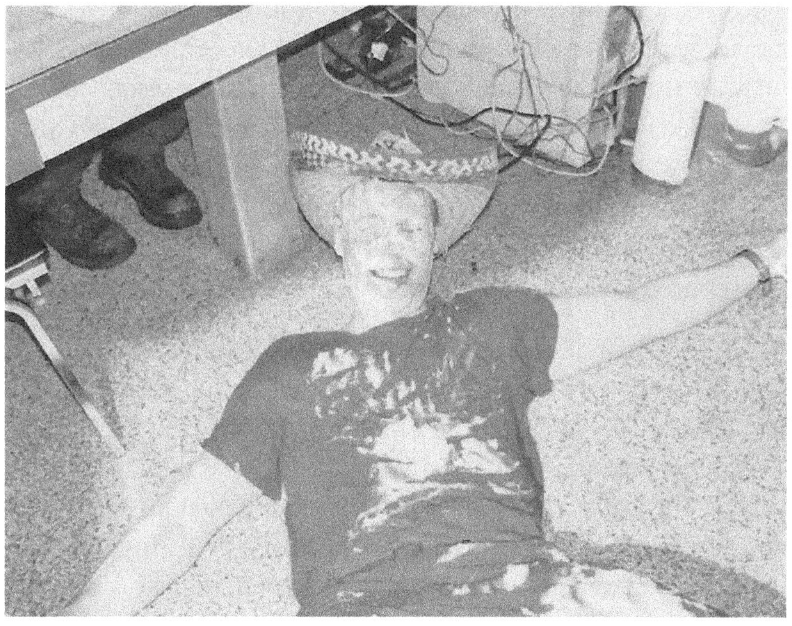

Sombrero going away party

Both John and Matt went on to have great careers in and out of the Coast Guard. As of 2021 John is a warrant officer and Matt got out of the service and manages a grocery store, making more money than he would have in the Coast Guard.

Getting back to my normal duties; as the DCA I was responsible for the overall firefighting, damage control, fueling, training, and stability of the cutter. To learn my trade I was sent to the Navy's damage control assistant, DCA, School in Newport, Rhode Island. DCA School reinforced what I had already experienced over the years being assigned to repair parties and firefighting teams on previous cutters.

During a casualty at sea such as a fire, collision, flooding, or attack, we would sound an alarm and go to general quarters, GQ. When this happened, I ran the total casualty response from a location known as damage control central, DCC. This location was in the aft part of the cutter, but if it was compromised, I had various fall back points containing all the ship's diagrams and communication systems I needed to direct a response.

In DCC I had a diagram of every system on the cutter along with a communications system comprised of two sound powered phone talkers. You can see the sound powered devices around the necks of my two phone talkers in the next photo. Basically, the devices worked like the old tin cans and a string set up you had as a kid- they needed no electronics to work. My phone talkers were in contact with the various repair lockers and the captain on the bridge. Through these phone talkers I received information on the casualty and was able to disseminate a plan of action. There were two repair lockers on

Dependable, which was typical of all cutters I was on. Repair 2 was the lead repair locker and usually handled issues forward of the cutter. And Repair 3 which was the secondary repair party and handled issues in the aft section of the cutter.

The following picture is in DCC and we are wearing our battle dress with flash gear. Left to right is Electrician Mate Third Class Dudley, me, and Operation Specialist Third Class Bevilacqua.

Damage Control Central

Let's say we had a fire. General Quarters would be sounded and everyone would man their GQ billets. Once I received the manned and ready from my repair lockers I would order the nearest repair locker to the fire to set the required fire boundaries, provide routes for the fire team to get to the fire, and identify any problems that may arise because of the location of the fire. Don't forget we were a military vessel and carry lots of ammo! As the repair locker set boundaries and fought the fire they would report back to me on their progress every step of the way. If the sound powered phone circuit would go out, written messages using a sort of damage control shorthand would be used via human runners. Redundancy was built in to everything we did.

EM3 Lesniak conducting Damage Control Training

Like every other cutter I had served on we trained in damage control several times a week in port and underway. At least one day a week was dedicated to all hands drills. Every evening the duty section trained in damage control. We had to be able to not only handle issues on our own cutter but also to go and assist other vessels. So we had to know what we were doing.

While underway it was my daily responsibility to calculate all the liquid loads and weights above the main deck and provide the captain with the stability curve for the cutter at the 1900 evening report. This curve basically told him what the cutter's maximum degree of roll would be before capsizing. This was obviously very important in rough seas. And stability was maintained by moving fuel around from tank to tank and filling or emptying ballast tanks with sea water.

As the electrical officer I was in charge of the five man electrical division. They had a chief, Electrician Mate Chief Terry Collins, running them so I really didn't have to manage them, I just signed paperwork and represented them at captain's mast!

Speaking of Terry, I had the honor of being part of his wedding ceremony. It was held at the chapel on the base there in Cape May. It was pretty cool because we did one of those sword arches that you see in the movies when the bride and groom exit the church and walk under the arch of swords. Thinking about it, it was probably the only time I ever used my sword besides learning movements with it at officer indoctrination school. And I'm glad my only official function with my sword was for Terry's wedding.

Me and MKC Shea at Terry's wedding

I would make eight patrols on *Dependable*; three in the North Atlantic and five in the Caribbean Sea. I experienced more of the operational tempo aboard *Dependable* than on other cutters. I say this because as a machinery technician I spent most of my time working on mechanical issues and wasn't really concerned with what the cutter was getting into. Occasional I would get called upon to do boardings as a boarding officer, as an engineer to go over and fix a mechanical problem on another boat, or to try and figure out if there was something suspicious with the machinery layout hiding a secret compartment. And I really never had a sense of the overall operational picture that the cutter was involved in. My concerns were keeping machinery operational and the next port call. But now I wasn't tied down to an engineering space. I had free movement all over the cutter and I was privy to all that was discussed in the wardroom where the captain and executive officer spoke freely about what was going on.

Administering the oath of re-enlistment to FS2 Pete McAndrew

Another cool thing about being an officer, besides having a cool sword I could use at weddings, was re-enlisting people. I hadn't been an officer very long and had already done one at my previous unit in Boston. I think guys sought me out because it was a novelty to be re-enlisted by someone who was once a master chief. Because junior officers typically didn't do re-enlistments at bigger commands where higher ranking officers were plentiful. In the previous picture I'm re-enlisting FS2 Pete McAndrew in the cutters wardroom.

I think I was afforded the opportunity to interact more with this crew than any other cutter I had served on. Even though I was the command master chief on a 378 foot cutter previously it had a larger crew of 160, which was tough to get to know everyone. Plus my primary job of managing the engine room kept me from getting to know and interact with everyone really well. As a collateral duty I never really got to do my command master chief job as well as I had wished and the majority of my time was spent on engine room issues.

Here on *Dependable* my responsibilities as the DCA actually had me working with every division and constantly walking the cutter checking on firefighting equipment, doors, hatches, and all sorts of material condition classifications. And these items were maintained by crewmembers in each division. This enabled me to interact with everyone. Because I had a casual relationship with everyone it got to a point where I was frequently asked questions about the length of the patrol, how many days were we legally allowed to be held out at sea, do you ever have to go back out to sea again if your ship sinks, and all the important things a sailor, or cutterman, has on their mind!

With all these question I was being asked it gave me the idea to start a sort of question and answer forum and I ran the idea by the captain. He said "sure, why not. Just don't let it get out of hand" and I started the "Ask the JG" forum in which you could submit your question to me, and I would post the answer on our bulletin board. I would like to think it was a good way for the crew to sort of vent issues without getting into trouble. Because it seemed like most of the questions revolved around current events on the cutter. Such as “can the cooks legally serve us chicken 5 days a week” and “aren’t we legally supposed to pull into port after being underway for so many days.” But I also enjoyed question such as “what’s the history behind Coast Guard mascots at sea.” These questions kept me busy, and I think provided a nice morale boost for the crew.

The guys I really enjoyed spending time with were the cooks. And I’m not just saying that because I have a sweet tooth and always hungry. Their rating specialty was Food Specialist, FS. It was previously Substance Specialist, SS, for the longest time but they had recently decided to change the name. The cooks were the guys that had to get a meal out to the crew no matter what. And in rough seas, which were normal out in the North Atlantic where we predominantly patrol, that isn’t easy. In this following picture you can see food and pans that got tossed to the deck after hitting some rough seas.

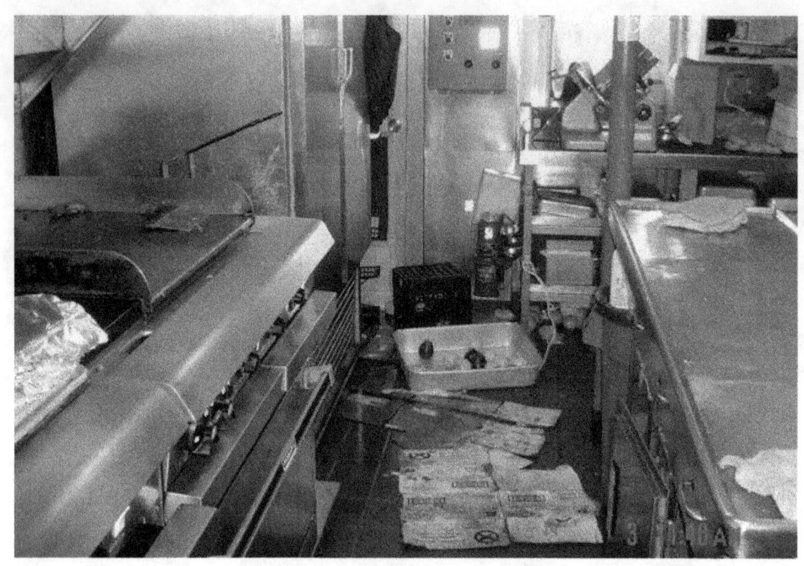

Galley after rough seas

But they did it plus baked in the evenings so we could have something sweet to snack on. And like everyone else they had collateral duties to also attend to and when the GQ alarm sounded they ran to their billet or when Flight Quarters was conducted they also had to break away and do that.

In the following pictures you can see them baking cookies and cakes. Some of this they had to do with their normal cooking duties but a lot of it they did as extras on their own time because they wanted to give their shipmates something special with their meal or as a snack.

The FS3 Jones & FS3 Clarke baking cookies

FS2 McAndrew with a banana nut cake

And I guess to sum it up I would have to credit both my years of experience and the freedom of being an officer that enabled me to compile this book. For most of my career, as I have previously mention I was concerned with my own responsibilities and the next port call bar. But now I had the time and opportunity to document what was happening day to day. Along with whatever I was writing for the "Ask the JG" forum I was also going around and taking pictures of what people were doing every day and posting them on the bulletin board. I would also send them to family so they knew what we were up to. Plus, when we conducted migrant interdiction operations my billet was to video tape and photograph the process in case something happened, and we needed to document the event. Add the fact that I actually saved a lot of this information for almost 20 years is what enabled me to compile such a detailing of events.

So, let's get underway.

2003

We slipped our berth from Coast Guard Base Boston at 1700 on the 6th of June and headed back into the Atlantic Ocean. I had just reported aboard the previous day and the crews patrol break was over, it was time to go to work.

On the 9th we were on patrol in the Atlantic Ocean, in D1's operational area, when were diverted to the disabled fishing vessel *M/V Rachel T* who was near Cashes Ledge about 90 miles off the coast of Maine. The *M/V Rachel T* had sustained an engine room fire and was disabled at sea. We took her under tow and brought her into Portland, Maine where the tow was transferred to a commercial tug to take her into port.

Usually when we towed a vessel in, we rarely brought it all the way into port. A Coast Guard small boat from the local station or commercial tug would come out into the harbor and relieve us of it and we would head back out to deep water.

19th of June, we received a search and rescue, SAR call that a fisherman had fallen overboard from the Canadian fishing vessel *M/V Atlantic leader,* about 50 miles from our location. He was part of a six man crew and had fallen overboard at about 2300. We came up to a full bell of about 18 knots and headed toward the SAR area located about 150 nautical miles east of Cape Cod. We arrived on scene just as I was getting off of the engine room watch at 0330. I stayed awake along with the rest of the crew searching the waters and horizon without luck. The water temperature was about 50 degrees, which would give the missing fisherman about an hour of survival time in the water without a protective suit. We never did find him.

27th of June, we escorted a fishing vessel deemed un-sea worthy into port. The boarding team, while conducting normal fisheries inspections, found that the vessel had numerous fuel and oil leaks in the engine room and no fire suppression equipment aboard. So we took the vessel in tow and headed for the nearest port. As we were dropping the fishing boat off we received a distress call from another vessel about 30 miles away taking on water. So, we headed out to her at the best speed we could make, which in the sloppy sea state was about 15-16 knots per hour. We arrived on scene with the disabled vessel two hours later, at 1325, and sent our rescue and assistance, R&A, team over. A Coast Guard HH-65 helicopter had arrived earlier and lowered down a P-1 dewatering pump to the boat that helped out with their water issue until we arrived. Our R&A team got the vessel's batteries charged and they were able to start their engine. The *Sanibel*, a 110' cutter, arrived on scene to assist. We left the fishing vessel with the *Sanibel* and continued on patrol at 1545.

Coast Guard HH-65 helicopters were the work horse helicopter at the time and had a good range. They could make it 150 miles or so off the coast and be able to return to their base. So they could usually get to a vessel in distress faster than we could and drop off a pump or hoist people to safety. With our flight deck we could extend their range by enabling them to fly out to us, refuel, and continue on.

8th of July, one of the female ensigns aboard had her appendix burst. It couldn't have happened at a worse time because we were way out to sea and the weather was really bad with a high sea state and driving winds. We were hoping she would stabilize as we worked our way closer to shore and within helicopter range, but her condition had worsened. So, with our options dwindling the officer's wardroom was readied in case an emergency operation had to be performed. The wardroom is actually designed as a make shift operating room, complete with operating lights and table. This is mostly due to the cutters military characteristics. Luckily the sea state calmed and a helicopter was able to make it out to us and pick up the ensign. Since we didn't have a doctor on board, only a corpsman, I'm sure the ensign was relieved she didn't have to hang around and have him perform his first ever appendectomy!

9th of July, we were diverted to the 44 foot sail boat *Alegria* that was about 170 miles south of Nantucket, Massachusetts to assist with a female crewmember suffering from severe abdominal pain. I mean what were the odds of two female sailors suffering from abdominal pain in two days! The *Alegia* had a crew of 4 and was sailing from Bermuda to Newport, Rhode Island when one of the female crewmembers became ill.

They decided to continue on thinking she was just seasick. Eventually they made radio contact with a passing cruise ship and the doctor on board told them to call the Coast Guard for medical assistance. Since we were nearby we were directed to transfer the sick crewmember aboard *Dependable* and then take her into port. We started heading towards the *Alegia* in 45 mph winds and 15 to 20 foot seas. While in route to the vessel a Coast Guard helicopter arrived on scene and picked up the sick crewmember and flew her ashore to a medical facility. The doctor who treated her said it was the worst case of appendicitis that he has ever seen anyone survive.[3]

And that was typically how a search and rescue case would unfold for us. With the cutter plowing through rough seas for hours on end getting the crap beat out of us just to have a helicopter pluck the injured person, or entire crew, up before we could get there. Then they would fly back to their air base to awaiting news crews who would fawn over them, report how heroic they were, and plaster them all over the 6 o'clock news. And just maybe there would be a slight mention of some cutter that assisted in the operation. Meanwhile we would still be out there battling the rough seas looking for anyone else who might not be accounted for and dealing with the vessel, if it was still even afloat.

16th of July, I was awakened at 0215 for a SAR call 40 miles from our location involving five people who were in the water after their fishing vessel sank. We had been idling around in dense fog for the past several days so the transit to the distressed vessel took longer than normal with us slowly feeling our way through the fog and blasting our fog horn. We arrived at the last known location of the fishing vessel *M/V Iiha*

Bravo, which was 69 miles east of Chatham, Massachusetts. A Coast Guard helicopter had arrived before us and had hoisted the five crewmembers into the helicopter flying them to safety. There was no sign of *M/V Iiha Bravo* when we arrived on scene and she was presumed to have sunk. We noodled around for a while looking for debris but didn't find anything.

We conducted normal fisheries boardings until we returned to our home port in Cape May on the 28th of July for our scheduled in port period. During the patrol we conducted more than 70 boardings and participated in 7 multi-unit search and rescue cases.

On the 13th of September we shoved off for the Caribbean Sea where we would be patrolling off the coast of Honduras. Our patrols were usually broken up this way. Doing a fisheries patrol and then a drug and migrant patrol. I think it kept us proficient at both because they were totally different as you will see.

On the 15th we conducted helicopter operations and embarked our HH-65 helicopter for the patrol. We typically didn't carry one while doing fisheries patrols up north because the weather was to rough and we didn't have a hanger. We also didn't have a helicopter permanently assigned to us. Instead, a different helicopter and crew was assigned to us from a rotating pool if our patrol required one. Usually on cutters 270' and greater that had helicopter hangers the helicopter would join the cutter soon after it left its homeport. But because *Dependable* only had a flight deck, and no hanger, our helo usually waited to join us at the last possible minute. In this case it was South Florida.

17th of September, we stopped in Key West, Florida for fuel and stores and loaded up on personal items because this would be our last U.S port call for a while. We had a fireman, FN, who was restricted to the cutter due to punishment he received at captain's mast and he needed to go to the exchange to purchase toiletries. He wasn't well liked and no one wanted to be bothered with him. It wasn't unusual for guys to get into trouble. Being young, carefree, booze and port calls, sort of led to getting into trouble. But not caring about your job and shipmates got you alienated really quickly.

I was heading that way and said I would escort him. I was always willing to give these young guys the benefit of the doubt; probably because I saw a little bit of myself in them. Heck, I did just as many stupid things as they did when I first came in the service. If the Coast Guard hadn't decided to invest some time in me after busting me for dope, where would I be now? In my position I felt obligated to try and mentor guys like this. Sometimes I got through to them and sometimes I didn't. But at least I tried. The exchange was about a mile walk through the Coast Guard Base and onto the adjacent Navy Base. I struck up a conversation and we chit chatted about general stuff along the way. As we came out of the exchange, I noticed a paper machine across the street and walked over to it to get a paper, not thinking about leaving the fireman. When I turned around, he was gone! I was thinking, no way! Where could he have bolted to so fast? I hurried around the corner of the exchange and there he was smoking a cigarette. I guess I should have had a little more faith in the kid. And I was relieved I didn't have to go back to the cutter without him and explain that he gave me the slip.

It reminded me of another story that a salty guy I used to work with named Gil Parks had told me. Gil was a Machinist Mate Chief back in the 50’s, 60's and 70’s and rode the old steam boiler powered cutters. And when I say he rode them I mean he would do tours on them back to back. Now this story was from sometime in the 50s or early 60s. He said that he was detailed to drive a prisoner up to Fort Jay, the Army prison on Governors Island, New York. Which eventually became a Coast Guard base. On the drive up the prisoner said he had to take a leak, so Gil pulled the car over to the side of the road and let the prisoner out to do his business. Gil said he sat there in the driver’s seat and lit a cigarette. He looked into his rear view mirror to check on the prisoner, who should have been in the bushes taking a leak. Damn if the guy wasn’t about 50 feet from the vehicle and hitch hiking! Before Gil could get out of the vehicle and grab his prisoner, the first car that came along pulled over and picked him up, probably because he was in uniform. Gil tried to wave the car down but the car sped off. Gil hopped back into his vehicle and began to chase them. He caught up to the car and honked and waved until the car with the prisoner pulled over. He eventually got the prisoner back into custody and off he went to Fort Jay with him!

19th of September, we departed Key West and had just hit the sea buoy when we were given custody of a Cuban migrant who was picked up off of Key West. The local small boat station had intercepted him and we were tasked with transporting him to Guantanamo Bay, Cuba or what we called Gitmo. The policy on Cuban migrants was that as long as they didn’t touch U.S. soil they had to be taken back to Cuba. If they touched U.S. soil they had to be held in the U.S.

On the 22nd we pulled in to Gitmo to disembark the Cuban and took the opportunity to top off on fuel and stores.

26th of September, I was awakened by the announcement on the intercom system for my roommate, Ensign Jack Souders, to report to the boat deck ASAP! I looked at my clock and it was around 0500. With all the pipes over the PA system and Jack banging around I decided to get up. It wasn't really dark in the room because all the berthing areas typically kept their red lights on at night. This allowed the watch stander coming to wake you up the ability to easily find your rack without stumbling around in the dark. And most importantly it kept your night vision, which was especially useful if you were going up to the bridge on watch.

Navy frigate assisting with drug bust

I headed one deck up to the wardroom to get a coffee. Scuttlebutt in the wardroom was that we had intercepted a "go-fast" boat. The term "go-fast" was used to describe a drug running speed boat. I pulled back the curtains on the wardroom porthole and there, about 200 yards off the port beam, was this huge Navy frigate, the *USS Crommelin FFG-37*, with a helicopter hovering around it. I decided to get a better look and took my coffee out to the weather deck. There, about 50 yards in front of *Dependable*, was the "go-fast" surrounded by our two small boats and one from the Navy frigate. Later that evening as I was coming off of my four hour engine room watch, I stepped out onto the weather deck to get some fresh air. It was dark out, a thunderstorm was brewing on the horizon and you could see the lightning in the distance. The fantail light was on which is unusual at night because we steam in a darkened ship status to avoid detection. And like I previously mentioned if a light was on it was a red light.

All vessels are required by international navigation law to have navigation lights on that indicate their size and port and starboard sides. When we conducted drug and migrant operations we usually ran with these lights extinguished so you could not see us. We considered it a military operation and didn't have to obey international law.

I walked back toward the light and saw we had prisoners from the "go-fast" and they were lounging around under the watchful eye of our armed crew. As I walked back towards the forward part of the cutter, I could see our small boat was alongside of the cutter disembarking passengers. A few people got off and they headed toward me making their way back aft to the fantail. More “go-fast” prisoners. I thought “wow we must have made another bust!”

The “go-fast” is hard to catch. They run at night, at high speeds with no lights, and during the day they stop and throw a big blue tarp over themselves so you can’t see them. They also have what are known as “mother ships” stationed on their drug running route that provide them with fuel. It’s quite the logistical set up and of course huge money.

Captured “go-fast” boat running drugs

With all that going on we still made time to take care of some other official business. There are several centuries old naval traditions that are observed to this day that revolve around sailing through certain areas of the seven seas. The open water is King Neptune’s kingdom, and all sailors must pay respect to his highness. If not, you’re sure to be sunk and join Davy

Jones at the bottom of the sea! Crossing into the Caribbean Sea, also known as the Spanish Main, is one of those areas deemed sacred by King Neptune. So, when we crossed into the Spanish Main we held a ceremony to appease King Neptune. It turned Scavengers, Goldbricks and other Scurvy - those who have never crossed into the Spanish Main - into sailors - those who have. It does not matter how long you have been at sea or your rank; if you have never crossed you're initiated.

King Neptune

So, on the 28th of September we stopped what we were doing and held the initiation – oops I meant to say ceremony. It turned Scurvy into the brave and brazen who were entitled by the laws of the sea to brag, swagger, spit into the wind, and otherwise act in the manner of the crews of the *Jolly Roger*. I had already sailed the Spanish Main on three other cutters so I didn't feel the wrath of King Neptune and his court.

Spanish Main Ceremony

15th of October, we had pulled into Guatemala for a few hours for fuel and to remove garbage. And we desperately needed to get rid of the garbage we had because it was piled up everywhere. The stink of it seemed to be everywhere on the cutter as it cooked in the Caribbean heat. For some reason garbage was never really a problem on the other cutter I served aboard. But on *Dependable* the trash room filled up fast and it always seemed to be piled up on the fantail and along the aft air castles. We had a garbage compactor but it really didn't seem to keep up with demand.

In hindsight I think it had a lot to do with newer regulations restricting what could be thrown overboard. On my previous cutters just about everything except for plastic went over the fantail – cardboard, metal, and food. On *Dependable* only food went over. Everything else was held on board until we reached a port to dispose of it.

Guatemala would have been a nice liberty port call but we were scheduled to pull into Roatan, Honduras - which we did on the 17th. There weren't any services there, such as fuel, so we had to get it in Guatemala, which was not far away. The interesting thing about both places, besides being really poor, was the level of firearms the guards walked around with. It was routine to see guards carrying army grade machine guns around any business of importance like a bank or the fuel piers.

Fuel pier in Guatemala with armed guards and Ensign Jack Souders

When we were in Roatan the local police chief came aboard and I had some coffee with him in the wardroom. He really wanted a new holster for his pistol, but we were not authorized to provide him with one. We did trade coffee with him thinking we were making out by giving him Maxwell House for some fresh South American coffee. But it tasted awful, like sawdust. I was really disappointed because I was a really big coffee drinker.

The captain and the executive officer were both really into bike riding and went out on a mountain bike ride one day. The XO ended up taking a really nasty fall and dislocating his shoulder. Since this was way beyond what our corpsman could handle, he was taken to the local hospital. But after arriving there it was apparent that our corpsman probably knew more about setting the shoulder and had a more sterile environment to work in than the local hospital. The XO's condition was pretty bad, and he was eventually airlifted to a U.S facility.

It reminded me of when the XO of another cutter I was on went down on the bridge. It was on the *Tampa,* and we were patrolling south of Puerto Rico. We were conducting a boarding on a suspect vessel, and I was back on the fantail taking a break from the engine room and watching the boarding unfold. All of a sudden *Tampa's* horn started blasting away, which was totally unusual. Next thing I know Flight Quarters is being piped and people are running all over the place. Once I got settled into my Flight Quarters billet in the engine room I finally learned that the XO had had some sort of heart attack or heart problem while standing on the bridge and collapsed. The bridge watch immediately sounded the horn to get the boarding team and small boat back and set flight quarters to medivac the

XO. We had a helo on board, so he was quickly transferred to Puerto Rico. He survived and I later served with him at another unit. But when he went down it scared the crap out of us!

After leaving Roatan we began patrolling in the northern part of the Caribbean Sea. Things were slowing down and we were getting ready to pull into Gitmo for a patrol break. Gitmo wasn't usually looked upon favorably as a port call. Besides being notorious for holding terrorist it's just a typical Navy Base, and that's it. Most of the base is off limits and you can't leave it and venture into Cuba, unless you want to walk through an active mine field. If I had to pick some positives it would be the club, Post Exchange, and the McDonald's restaurant.

Gitmo used to be a pretty active base and I had been pulling in to it for over 15 years. But over the years they have moved a lot of the naval units out and the base pales in comparison to what it used to be. Probably my most memorable time there was when I was on *Seneca* and the battleship *Iowa* was in there with us for a week conducting shipboard training. The size of her was impressive. A few weeks later when we were patrolling off of Puerto Rico and *Iowa* was performing gunnery exercises off of Vieques Island her #2 gun turret exploded, killing 47 of her crew. We stood by not knowing if we would be called upon to assist. But eventually it turned into a Navy matter and we continued on patrol.

Early in the morning on the 24th of October we were informed that there was a "go-fast" operating in our area and we were to divert and intercept it. We launched our helicopter and began the search. The "go-fast" was spotted by the helicopter and the chase was on. Needing to refuel the helicopter returned to the cutter and was quickly redeployed to track the "go-fast." As we neared the "go-fast" our small boats were deployed, and we closed in on her with the helicopter in the air and small boats in the water.

As we closed in, the crew of the "go-fast" started to throw bales of drugs overboard to lighten their load so they could make better speed. Each bale weighed about 75 pounds and was causing a considerable drag on their boat. Eventually the helicopter was once again running low on fuel and the small boats were slowing to recover the jettisoned bales of drugs. We needed to break off pursuit, recover the helicopter, and relieve the small boats of their cargo. Despite all our efforts the "go-fast" was able to escape. But we had recovered over a ton and a half of cocaine with an estimated street value of $75 to $100 million dollars!

In the following pictures you can see the bales of cocaine our small boat crew was having to stop and pick up as they were jettisoned from the "go-fast." And in the following picture the recovered cocaine stacked up on the forecastle of the cutter.

Recovered cocaine from go fast

Recovered cocaine

That was it for this patrol. We out-chopped from our operational area and started heading for home. But heading home didn't mean we were free from any responsibilities. It was always part holding our breath that nothing would cause us to divert, such as a SAR case, and part excitement that we were heading home. And on this occasion, we quietly sailed north, flew the helo off as we passed Florida, and eventually rounded the sea buoy on the 1st of November and returned home to our moorings in Cape May.

Law enforcement was a huge part of what we did. Whether it was drugs, fishing law, illegal migrant interdiction, we did it all. And law enforcement was a collateral duty just like everything else. There was no law enforcement rating and the responsibility was shared by the crew, so we all had to be trained to carry a weapon and pepper spray. To carry the service issue 9mm Berretta pistol you had to qualify at the range and the judgmental course. The range was obviously live firing the weapon at a target. Judgmental was done by standing in front of a TV screen with your weapon holstered as different scenarios played out in front of you. At some point you would have to decide whether to draw and fire, or to hold your fire. As an example, you would be entering a space on a fishing vessel and there would be a person in front of you holding a knife. At some point he may come at you with it to stab you, or he might just drop it. You had to decide if the situation required you to draw your weapon and command him to drop it. Sometimes he would come at you and sometimes he wouldn't. You had to know when to fire and when to hold fire. It seems simple enough but spaces on vessels are small and tight. So, you have a split second to make a decision. And it better be the correct one.

Usually, the small boat crew that carried the boarding team over to a vessel were not armed. But if the vessel was known to be a threat, like a "go fast," the small boat engineer carried a shot gun.

To carry pepper spray you had to actually be sprayed with it so you would know how effective it was. I can tell you it burns like hell and for a long time. It always sort of amused me that to carry pepper spray we had to be sprayed with it but to carry a pistol we didn't have to be shot with it.

FN McDonald Being pepper sprayed and fighting off an attack from MK3 Diaz

There was also training on defending yourself after being pepper sprayed – which could happen if you got down wind when spraying it at a suspect. And on subduing and handcuffing a suspect.

In all my years of doing boardings I never had to draw my weapon or subdue anyone. Which is sort of surprising because I have boarded vessels carrying drugs and even Iraqi vessels violating United Nations sanction in the Persian Gulf. Usually as bad as it would get was having to stay the night on a nasty, dirty, stinky vessel after we had seized it. I was surprised how no one resisted. Especially with the jail time the drug runners were facing. They would be compliant and seemed matter of fact about it. Probably the most contentious people I boarded were fishing boats and illegal migrants. For some reason the fisherman seemed to have an attitude. Probably because they were sick and tired of us bothering them all the time. But they never complained when we came to pluck them out of the water. And of course, the illegal migrants were understandably not happy we were there to end their dreams of a better life.

16th of December, we hauled in the brow and set a course out into the Atlantic Ocean on what would only be a seven day patrol for me. The rest of the crew would be at sea for Christmas and New Year's.

20th of December, while conducting fisheries inspections off the coast of Long Island, in D1's operation area, we were dispatched to a SAR case involving a sailing vessel. The weather had gotten really nasty with high winds, 13-15 foot sea state, and 38 degree water. The fishing fleet had huddled together for safety, and we had closed in to keep an eye on them. One of the fishing vessels became unstable and we began to escort it into New York harbor. This wasn't unusual because if a fishing vessel returns to port without catching its quota its losing money. So fishing vessels stay out in bad weather trying to maximize their catch. But the more they catch, the more unstable they become in rough seas. That's why we stay close to them when it gets snotty out, because we know they are going to need our help. It's a dangerous business.

On the way into New York we received a distress call from the 40 foot wooden sail boat *Gaucho*, which was near our location. I was told the vessel had an interesting history and it was once owned by President Franklin D. Roosevelt. But I have never been able to verify that.

We diverted from our escort and headed toward the location of the *Gaucho,* which was about 80 miles off of Shinnecock Inlet. When we arrived at 0200 in the morning we found the *Gaucho* was out of control and floundering in the wild sea state. The story was that there was a crew of two, a husband and wife. The husband had fallen overboard attempting to fix something on deck and the wife, unable to help her husband, became hysterical and fled to the safety of the inside of the boat. A Coast Guard helicopter had beaten us to the scene and had plucked the husband out of the water and was taking him to Stoney Brook University Hospital. The weather was to rough, visibility to bad, and the wife too distraught to attempt to remove her from the vessel by helicopter. Plus, the mast of the sail boat was whipping around making a hoist from the boat impossible.

As we assessed the situation nothing was working in our favor. Initially we thought that we could wait the weather out and then get a boat in the water and rescue the wife. This plan was soon not viable when the weather and sea state put the boat in jeopardy of sinking. We had to somehow get our small boat in the water and get it over to the floundering sail boat before it sank taking the wife onboard down with it. After getting the cutter in the best leeward position to lower our small boat, we launched our ridged hull inflatable, RHI, along with a small boarding crew.

It took the RHI and her crew about 30 minutes to get to the *Gaucho* which was only several hundred yards away. Try after try failed to get the RHI alongside and a crewmember aboard. Finally, a heroic seaman, SN Benjamin Murphy, coaxed the RHI coxswain to get the RHI close enough for him to make a jumping attempt to get onto the *Gaucho*. As the RHI neared the *Gaucho* SN Murphy leaped aboard. If he had missed, he could have been crushed by the two boats slamming against each other. Once on board he located the wife huddled in the lower cabin and administered her comfort. He then managed to pull two more of our crewmembers on board directly from the RHI to assist him. Together they doused the sail and started the trolling motor, which would get the boat somewhat under control.

Once these tasks were accomplished and the situation stabilized they were able to transfer the wife to the RHI and then to the cutter. The three Coast Guardsmen stayed aboard to hook up the tow rope. The sea state was so bad that they had to stay aboard the *Gaucho* overnight until daylight when the towing attempt would be safer. Once the *Gaucho* was in tow we headed into Long Island, New Jersey.

I had the morning watch the next day and entered the wardroom for breakfast. The sea state was still rough and just walking to the wardroom was a challenge. While having my breakfast the wife from the *Gaucho* walked in and sat down. I started a conversation and we talked about her ordeal. She said she and her husband were taking the *Gaucho* south for the owner when they ran into the bad weather. Her husband had gone topside to take care of some rigging and had fallen over board. She did not know how to sail and could not get the boat to come about to try and pick up her husband. She dispatched the distress call and the Coast Guard helicopter came out and picked her husband out of the icy cold water.

They were both wearing the appropriate cold weather suits and she thought he would be okay once the Coast Guard helicopter picked him up. I reassured her that everything would be fine and we would get her reunited with her husband as soon as possible. She left the wardroom and the XO came in. I asked him if he had heard anything concerning the husband and he told me to keep it hush-hush but he had passed away on the helicopter in route to shore.

On the 23rd of December we pulled into the inlet at Sandy Hook, New Jersey and transferred the tow of the *Gaucho* to another Coast Guard cutter. I was getting off the cutter to attend a damage control school and escorted the wife ashore via a Coast Guard small boat from station Sandy Hook, NJ. By this time the captain had informed her that her husband had passed away, so the boat ride in was very somber. We sat next to each other in the warm cabin of the small boat but I didn't try to make small talk. I could tell she was just trying to keep herself composed, and it wasn't easy, so I just left her to her thoughts. When we arrived, I escorted her to the station's mess deck where her family had gathered. It was an emotional meeting as expected. I was relieved to be able to hand her over to the good care of her family.

The patrol was over for me, and I headed to Newport, Rhode Island to attend Surface Warfare Officer Damage Control Assistant School until the middle of February. After I graduated on the 19th of February, I caught up with *Dependable* back in Cape May.

But *Dependable* still had another search and rescue case after I left. On the 7th of January she was patrolling off of New York Harbor due to a heightened security level. This was just after the attack on 9/11 so that area was heavily patrol. She was dispatched to assist the 44' fishing vessel *West Head* out of Portland, Maine. She was in route to the fishing grounds when she decided to stop and anchor to ride out a storm that had blown in. While she was at anchor the storm got worse, and winds started gusting to 60 mph and the seas raised to 20 to 40 feet high. One of these waves smashed out the *West Head's* wheelhouse window destroying most of her electronic and navigation equipment.

The *West head's* captain decided after that it was probably better to ride the storm out instead of being anchored and they covered the broken window with wood and cut loose their anchor. Soon after that they spotted the tanker *Talisman* and signaled for help. The *Talisman* then radioed the Coast Guard for assistance and stood by the *West Head* keeping an eye on her.

Making a 26 hour dash in rough seas *Dependable* arrived on scene at 0445 the following day. They were now about 150 miles due east of Cape Ann, Massachusetts. The crew of the *West Head* were brought aboard *Dependable* where they were able to contact their families and let them know they were okay. Meanwhile the *Dependable's* crew took the *West Head* in tow and delivered it to the closest port of Provincetown, Massachusetts.[1]

2004

On the 14th of March 2004 we hauled in our mooring lines and headed south once again for the Caribbean Sea. This would turn out to be an action packed patrol with illegal migrants and drugs.

We made our transit into the Caribbean Sea and refueled in San Juan, Puerto Rico on the 18th before slipping out into our operating area on the 19th. Things were initially pretty normal and we boarded numerous vessels without any issue. We also conducted drills and kept busy with damage control training.

There was always something going on during the day and into the evening with work, standing watch, and morale movies and events. Several movies were played over the entertainment system which ran through TV's every evening and most of the crew watched them on the crews mess, chief's mess, and wardroom. The officers even had TV's mounted in our staterooms and could watch them there. There were also morale events such as beard growing contests, arm wrestling competitions, and morale casino night.

When a crew stays busy and occupied it keeps their minds off of other things, like home and family - and underway morale is much better.

Just an interesting note on this following picture of morale casino night. The guy sitting in the center is LTJG Josh Burch and this was his first tour after graduating from the Coast Guard Academy. He would later go on to command *Dependable* 17 years later in 2021. And the guy to the right standing, EM3 Matt Lesniak, was the 2 time light weight arm wrestling champion aboard *Dependable.*

Morale casino night

After patrolling for 10 days we made it to our next port call in St. Thomas on the 29th of March. After a few days of liberty, we were heading back out on the 31st to see what was going on.

SN Hargis and his tuna

One of the nice things about sailing around in the Caribbean Sea was the great fishing. Although I wasn't that much into fishing, a lot of the crew were. So when we had some free time the captain would have fish call piped, which went like this – "Now fish call, down down all lines, up up all fish, now fish call." Everything was piped on a cutter. When to get up -

"Reveille reveille reveille, heave to and trice up, now 0630, now reveille." And when to go to bed – "Taps taps lights out, now 2200, now taps." But getting back to fishing, the previous picture is a nice tuna caught in between boardings.

3rd of April, we ran into our first yola - a small open boat - of Dominican migrants and they were not happy to see us. They were illegally headed to Puerto Rico with hopes of getting to the U.S. mainland from there. We easily followed the yola, it wasn't going that fast, and tried to stop it using our small boats and boarding teams. But when the small boats neared the yola the migrants started waving machetes and other weapons threatening our teams. When the small boats continued their approach the migrants then began to throw their shoes and personal belongings at the boarding team members!

We had just been trying out a new entangling device that was shot into the yola's propeller, fouling it and causing the boat to stop. This was a perfect opportunity for it and it was deployed by the boarding team and it worked, and the yola stopped. This, however, made the migrants even madder and we were basically at a standstill. So, we started to rig up the fire hoses on the forecastle thinking we would just blast them with water until they gave up.

Luckily there was a Dominican Navy vessel, the *Canopus*, in the area and it had come over to assist in the detention of the migrants. We gladly turned the situation over to them and were on our way.

Dominican Republic *Canopus* GC-107

It didn't take long to run into more migrants. The very next day while patrolling off the coast of Puerto Rico we intercepted a 30 foot yola carrying 121 illegals from the Dominican Republic off the coast of Borinquen, Puerto Rico. The yola also refused to stop and we followed it as it headed back to the Dominican Republic. Once in Dominican waters the yola was stopped and boarded by another Dominican naval vessel. Later on that day two more yolas were spotted off the coast of Desecheo Island, Puerto Rico. We stopped them and took 63 Dominicans into custody for illegally trying to enter Puerto Rico. We later took them into the Dominican Republic and repatriated them back to their home country.

6th of April, we were diverted to an incident involving a 23 foot yola carrying 38 illegals - 24 Dominicans and 14 Peruvians - off the coast of Punta Boqueron, Puerto Rico. The yola was stopped by another U.S. vessel and three illegals had jumped into the water. Unfortunately, the three drowned before we arrived on scene. The remaining 35 illegals were taken aboard our vessel. Later that day we offloaded the 35 migrants in Myaguez, Puerto Rico.

You just never know what's going to happen when you come upon these migrants and try to detain them. They have more than likely sold off everything they had to pay for the journey they are on and its America or bust. And when we come rolling up to put a stop to it you can understand why they fight back and do irrational things, like jump in the water not knowing how to swim.

6th and 7th of April, we intercepted another three yolas. The first was at around 2230 on the night of the 6th, again off the coast of Punta Boqueron. The 20 foot boat was carrying 18 Dominicans, 12 male and 6 females. In the meantime, two smaller Coast Guard cutters intercepted two more yolas containing 27 migrants on one and 54 on another. These illegals were transferred to us for a total of 134 illegals onboard *Dependable*. We later repatriated them back to the Dominican Republic on the 10th. Unlike the unruliness out at sea it was always an orderly affair and the transfer from our cutter to local authorities was uneventful. In the following picture you can see them walking away down the pier calm as can be. More than likely just to be let go to try the journey again.

Because they know the odds are with them. For every yola we stopped at sea there was an empty one found on the beaches of Puerto Rico with footprints in the sand leading inland.

Migrants being repatriated back to the Dominican Republic

Once we intercepted a yola and offload the migrants to *Dependable* we destroyed the yola. If there are any animals on board they're put down. Then it's burnt at sea or fired upon until it sinks. This is done for a couple of reasons. First, we just have no way of towing all these yolas behind us. Second, they are in such bad shape that they are not sea worthy. You can see a yola being burnt in the following picture. And in the next photo the Dominican migrants that came off of it on our flight deck.

Yola being burnt and sunk at sea

Dominican migrants on our flight deck

11th of April, we took on more illegal migrants and were pretty full at this point. We tried to make it as comfortable as possible for them and placed rubber matting down for them, so they didn't have to sit directly on the hot steel flight deck. Unfortunately, we didn't have a tent or anything to shade them. We set up a toilet and washing station on the flight deck so they didn't have to enter the cutter. For our protection we locked all the weather deck doors from the inside except the aft door on the fantail. This prevented them from hopefully gaining access to the cutter if they overtook the guards. As for food, they were fed a meal of red beans and rice three times a day. It seems sort of bland, but it was actually pretty tasty and it was a good meal for them.

As we were due to pull into San Juan, Puerto Rico that day for a patrol break we transferred the migrants to another 210' cutter operating with us, the *Resolute*, and pulled into San Juan.

18th of April, as the patrol was winding down, we were operating in the area of the Caribbean Sea known as the Mona Pass, which is between the Dominican Republic and Puerto Rico. While monitoring the suspected vessels list, we came across the *M/V El Conquistador* and boarded her. The boarding team found 4,840 pounds of marijuana in a concealed compartment worth over $10 million dollars. The drugs were seized and five people arrested. The next day we pulled into San Juan and turned over the drugs and prisoners to the Drug Enforcement Agency, DEA.

Our boarding team apprehending
M/V El Conquistador

Captured drug runners from the *M/V El Conquistador*

4,800 pounds of marijuana from *M/V El Conquistador*

We left San Juan and started our voyage north heading for Cape May and on the 24th of April we arrived back into homeport. It had been a long patrol. We had physically rescued 481 people from unsafe or sinking vessels from Haiti, Cuba, Peru, and the Dominican Republic. In all we had overseen the rescue of 899 migrants while supervising other Coast Guard assets and repatriated or transferred 752 people to the border patrol. Most were dehydrated and sea sick. One migrant was in such bad shape we had to medevac him off the cutter. This also included 35 people who were stranded on the uninhabited Mona Island.

Sailing the high seas had become a lot easier on me over the years and on *Dependable* we had all the comforts and conveniences of a modern ship. We had internet connectivity pretty much 24/7 and you could email when you wanted. There were communal computers and an email size limit, but now could email pictures. And as I previously mentioned I sent some out just about every day documenting the patrol. We also had satellite TV. And although its reception was dodgy, Captain Christian always tried to steer a good course for football games on Sunday.

Mail call was usually in every port of call and we only ran into problems when we got diverted from a scheduled stop and it had to try and find us. Cell phone use was forbidden when we got underway and they had to be turned off so that their signal could not be traced. Every once in a while if we were close to shore, and within cell phone range, the captain would lift the cell phone ban and let the crew make calls. This was always a morale booster!

Dependable hauled in her mooring lines on the 10th of July and set another course for the Caribbean Sea.

14th of July, we made a brief stop in Key West, Florida for fuel and to load our helicopter detachment from Air Station Houston, Texas. Their helo, 6571, landed, was tied down, and then we entered the D7 operating area. We would be patrolling off of Honduras and Nicaragua.

Heading out of Key West with our helo

22nd of July, we pulled into Gitmo for a two day break, fueled up, and took on stores. When we pulled in the place was dead. Besides the line handlers that helped us moor up there wasn't anyone else around. No fueling personnel, no one. It wasn't like there was always a band playing when we arrived, but there was always people there ready to take care of our fueling and food store needs. Since I was in charge of fueling I decided to walk down the pier and see if I could rustle someone up from the fueling department. So I walked down the 100 yard pier and headed towards some buildings where I could see some vehicles parked. As I got closer, I could see several very good looking ladies standing around with a group of other people. Curiosity got the best of me and I walked up and said, “Hello.” They replied “Hi, we’re Miss Universe, Miss USA, and Miss Teen and we’re doing a USO good will tour here on Gitmo.” I immediately forgot all about refueling!

Gitmo with Miss USA, Miss Teen, and Miss Universe

It was just so out of place for Gitmo. I had been pulling into Gitmo for a long time and never came across anything like this. As soon as it sunk in as to what was standing before me I knew the crew would love to meet them and I asked if they would come down the pier and aboard *Dependable*. They were most gracious and agreed. Needless to say they didn't need to be announced, because about half way down the pier the entire crew had noticed them! They were as gracious as could be and took the time to pose for lots of pictures.

After the USO good will group left we took on 19,080 gallons of diesel fuel, 433 gallons of aviation fuel, food stores, and then departed back into the operating area on the 24th of July.

29th of July, while patrolling about 30 miles off of Columbia we came across the Honduran flagged *M/V Miss Mery Hill*. We had received intelligence that the *M/V Miss Mery Hill* had left her homeport loaded with cocaine, so we had been looking for her.

She was heading on a northeasterly course when we picked her up and started to shadow her awaiting a Statement of No Objection, SNO, from Honduras. We could board U.S. flagged vessels at will but in order to board a foreign vessel we needed an SNO from the vessel's flagged country.

The SNO was granted, and we boarded her. The seas were calm, the weather very nice, and the boarding team had no problem making the transit and securing the vessel. Our boarding team however had been onboard for about 10 hours and had found nothing. The team was still hopeful because they were unable to account for space between some aft fuel tanks. These tanks hold thousands of gallons of fuel and we would need to pump the fuel out to be able to inspect inside them. Fuel, oil, and water tanks are tough to get dimensions on because they are built into the hull of the ship.

M/V Miss Mery Hill

It was determined that the boarding team would pump one tank into another to empty it for inspection. As the gas free engineer I would have to go over and make sure the emptied tank was free of explosive gasses and safe for entry. I received the word that the boarding team was ready and I made the transit over in the small boat with my gear, which consisted of an oxygen/explosive meter and an air tank - the type firefighters wear on their back.

As the small boat pulled alongside, I waited my turn and made the leap over to the *M/V Miss Mery Hill.* It was a typical Caribbean small fishing vessel about 80 feet long and rusting. I had to transit through the living area and you could see the boarding team had left no stone unturned. There were huge holes in the bulk heads, contents of every cabinet and drawer were emptied on the deck. The ceiling was torn down. The place was trashed with no sign of drugs. I certified several tanks gas free and let the boarding team go to work inspecting them. Since I wasn't one for hanging around half sea worthy vessels that stunk, I headed back to the *Dependable.*

The next day, the 30th, I was called back over to the *M/V Miss Mery Hill* to inspect several more tanks. I arrived onboard with my gear and made my way back aft. When I got back to the aft part of the boat I passed the six *M/V Miss Mery Hill* crewmembers who were under armed guard. I hopped down into a compartment to find one of the boarding team members, MK2 Shawn Hodgins who was on temporary duty with us, sweating up a storm while he was taking off the 20 or so huge bolts holding the cover on a fuel tank. I helped him remove the remaining bolts. When we finished removing the bolts we slid the 24 inch round cover to the side. We were looking into an empty small bathroom sized fuel tank with two more tank covers on the port and starboard sides of the tank. Now we were onto something!

The tank was empty, although the bulkheads were damp from condensation. You could clearly see that there were two more tank covers, one on the port side and one on the starboard side. MK2 Hodgins wanted to bolt right in but I put the brakes on him. If he were to just hop in there and there wasn't enough oxygen, he could just pass right out. If there were explosive gases he could blow us all up. I took a test and it showed good oxygen but borderline for explosive gases. I said I didn't feel good about him going in until it had been ventilated. Several of the boarding team members had come down and the adrenaline was really pumping. These guys had been here for over 20 hours tearing the place up with no luck and they wanted some justification for their destruction. Plus, we had been told there were definitely drugs on board.

MK2 Shawn Hodgins entering first fuel tank

Caught up in the moment I decided to let MK2 Hodgins go into the tank before ventilating it. I had gotten to know the him really well over the past month or so, standing watches in the engine room together, and knew that he was competent and would move quickly. And I would put my air tank on and be ready to go in and get him if something went wrong. Still, it was with some anxiety that I told him I would only give him time for one side, and he could pick. He picked the starboard side and went in. After about 20 minutes he had removed all the cover bolts and slid the cover plate off to reveal another empty compartment.

MK2 Hodgins working on the 2nd starboard tank cover

He wanted to start removing the port side cover bolts but I told him we would need to let the first compartment air out before I would let him spend another 20 minutes in there.

The airing out process would take several hours. I had no desire to sit over here in the heat waiting for this tank to air out so I told the boarding team that I was heading back over to *Dependable* and would come back in a few hours after the tank had time to vent. I hopped back into the small boat and told the coxswain to take me back over to the cutter. As I was jumping into the small boat my $100.00 flashlight slipped out of my coveralls and fell into the crystal clear water. I cursed as

I watched that $100.00 flashlight sinking to the bottom. And with those clear Caribbean waters it was a long cuss! It just didn't seem like things were going my way today.

When I got back over to the *Dependable,* I made my way up to the bridge to brief the captain on what had happened, minus the borderline explosive gases part. He was getting jittery because nothing had been found and he kept insisting the drugs were there. About the time I got up to the bridge everyone was really excited and asking me all about where the drugs were. Of course I was like "what drugs?" Come to find out the MK2 and the rest of the boarding team hadn't waited for the compartment to air out. As soon as I had hopped in the small boat they had crawled back into that empty tank and gone ahead and opened the port side cover. Inside the hidden compartment they discovered 64 bails containing 3,997 pounds of cocaine. It had a street value of $274 million dollars!

It was a huge bust and made the national if not world news. And for once when we arrived back in Cape May the local news came aboard and wanted to do a story about it.

You can see in the following picture EM3 John Dudley inside the first tank and looking into the port tank with bales of cocaine in it.

EM3 Dudley with the port side cover off and cocaine.

***M/V Miss Mery Hill* cocaine**

The vessel was confiscated and the six crewmembers taken into custody. While towing her toward Honduras she began to leak oil and was putting out a pretty good oil sheen on the water. At this point we were directed to transfer the tow, drugs, and prisoners to a U.S. Navy vessel. At some point during the tow they determined the *M/V Miss Mery Hill* wasn't sea worthy and she was sunk. The cocaine and crewmembers were later transferred to the U.S. Drug Enforcement Administration.

This seizure added another drug bust sticker to the side of *Dependable's* hull. This would be a snowflake indicating a cocaine seizure. Marijuana would be indicated by a marijuana leaf. Like human active duty members, cutters are awarded medals and citations for their accomplishments and they wear them on their hull proudly.

***Dependables'* awards**

We returned to Gitmo on the 4th of August for a quick refuel. No liberty, just fuel. After loading 18,097 gallons of diesel fuel we slipped back into the operational area.

10th of August, we were into hurricane season and category 4 Hurricane Charley caught us in the middle of the Caribbean Sea. We were lucky to be far enough east of it not to catch its full force. But the 50 mph winds and high seas still tossed us around. In the following picture you can see us plowing through the swells from Hurricane Charley. Our patrol break finally came on the 13th of August, and we pulled into the island of Aruba for liberty and took on 16,979 gallons of diesel fuel.

Headed to Aruba

Once we had the fuel on board and liberty was granted I went to take care of some paperwork and change into civilian clothes. I walked off the brow and into town and came across a couple of crewmembers walking down the street. I said "whatcha up to?" They replied "we just got back from the whore house." I said "shit we've only been on liberty for 30 minutes." They said "sir, it doesn't take long when you've been underway a few weeks."

As we set sail from Aruba we had another storm barreling down on us, tropical storm Earl. He was dishing out 50 mph winds and high seas right behind us. The storm was moving at 22 knots and we were doing 14 knots. He eventually overtook us and we were once again tossed around by the wind and the waves. It was a rough ride and there were a lot of sea sick Coasties!

But the weather eventually calmed down and we continued with boardings and even took some time to relax and shoot skeet off of the flight deck.

28th of August, as we were wrapping up the patrol we made one more stop in Gitmo for the usual fuel, food and an overnight stop. While we were having a morale cookout at the local beach the Washington Redskin cheerleaders showed up doing a USO tour! They were great ambassadors and took a bunch of pictures with the crew on the beach. Gitmo had changed. When we left Guantanamo it was the end of the patrol for us. We out-chopped and began heading north. And on the 7th of September we returned home to our moorings in Cape May where we remained until mid-October.

12th of October, *Dependable* once again slipped her moorings and entered the Atlantic Ocean on a D1 fisheries patrol. You can see our pier, next to *Vigorous,* in the following picture along with a small boat from the Coast Guard Station escorting us out. From here it was a short 10 to 15 minutes of steaming out of the Cape May Inlet and we were in the Atlantic Ocean and in the operating area.

Pulling away from our berth in Cape May NJ. That's the *Vigorous* in the background.

The patrol started off with a main engine casualty, which required us to pull into Portsmouth, Virginia for repairs almost immediately after getting underway on the 15th. It was reminiscent for me because I had done a tour on the *Tampa*, a 270' cutter, out of Portsmouth. Seeing the 270's moored up brought back fond memories of sailing on her as the main propulsion chief. You can see three moored up in the following photo.

Pulling into Portsmouth, VA Coast Guard Base

After a day of repairs and topping off with fuel we steamed back into the Atlantic Ocean on the 17th and headed north.

When we weren't boarding vessels or out on a search & rescue case we took the time to catch up on qualifications that had to be maintained. These included General Quarters drills on firefighting, collision at sea, enemy attack, and so on. One of these as I had previously mentioned was gun qualifications. Well, if you remember at the beginning of this book I mentioned that everyone had collateral duties, including the cooks. The following picture taken on the 28th of October is of the 25mm gun crew firing at a target, usually a drum we tossed into the water. Under the supervision of a gunners mate, GM, the cooks are actually firing the gun and handling the ammunition.

Firing the 25mm cannon

We pulled into Boston for a patrol break at the end of October. After two days rest we were once again making our way out of Boston Harbor on the 1st of November.

Just after leaving Boston on the 3rd of November we were directed to a fishing vessel in distress. We came up to a full bell and plotted a course to the vessel. It was reported that the vessel had sunk and the crew of 4 were in the water, so time was critical. Anyone who has been up to New England in November knows that the water is frigid and exposure can cause death in just minutes. But as usual a Coast Guard helicopter had beaten us to the scene and initially reported that there was no boat nor sign of the crew, just an empty raft. Then we heard that the helo had plucked the 4 crewmen out of the water. We arrived on scene about 25 minutes after the helicopter and retrieved the fishing vessels empty inflatable raft.

The crew was lucky because when these fishing vessels would go down they would go down fast, usually because they were overloaded with fish. When going down so fast they would suck the crew right down with the boat.

You can see what's left of the inflatable raft in the following picture. Pictured left to right are; Boatswain Mate Third Class Breeden, Boatswain Mate Third Class Dahl, Food Specialist Chief Sambenadetto, Seaman Murphy, and Boatswain Mate Chief Ryan.

FSC Rich Sambenedetto, or as we called him "Chief Sam" would be awarded the Douglas A. Munro Award for Leadership in 2005. If you think about it, he went up against around 30 thousand other coasties and won. That's a pretty big accomplishment. If you're unfamiliar with Douglas Munro he is the only Coast Guard Medal of Honor recipient. He earned it posthumously after evacuating Marines off of a beach at Guadalcanal during World War II.

Fishing vessels raft.

On the 13th it was back into Boston again for fuel and stores. These little stops were just long enough for us to take on food, fuel, enjoy a few hours of liberty and catch our breath. I had been stationed in Boston several times and knew my way around the city so it was always fun. And on the 15th we were back to conducting fisheries inspections.

We knew we would be spending Thanksgiving underway so we were looking forward to the feast the cooks would be putting out. Obviously, we would have preferred to spend the holiday with our families. But our shipmates were also our family, so it wasn't that bad.

FS3 Clarke baking pies for Thanksgiving Dinner

We were just settling in for our Thanksgiving dinner bobbing around off of New England when we were dispatch to what would be a pretty interesting rescue & assistance case. The 81' *Raw Faith* out of Rockland, Maine was disabled and floundering in rough seas about 50 miles off the coast. When we arrived on scene Thanksgiving Day I am quoted as saying "she looked like a pirate ship" in an email. And being built to look like a 16th century galleon with three sail masts she wasn't too far off from my description. She was built to take handicapped children out for sailing adventures. Their plan was to build the vessel with large flat decks to accommodate up to 30 children and their families.

The *Raw Faith* was headed to New Jersey to pick up more passengers before heading to Florida for the winter. But soon after setting sail she ran into bad weather, her rudder jammed, and about 6’ of her foremast had sheared off taking with it the radio antennas and navigation lights. When we arrived it was rough out and foggy as pea soup. Our initial boarding team found the crew cold and wet. So we brought 4 of the 6 crew members aboard *Dependable.* The other 2 remained aboard *Raw Faith* helping us hook up the tow line. But this turned out to be harder than expected because they didn’t have a standard towing bit installed on their bow.

At about 1800 we sent over our damage controlman first class, DC1, and electricians mate second class, EM2, to set up a dewatering pump because we found out that the *Raw Faith* was taking on water. It didn’t work out well and the DC1 got violently sea sick soon after getting aboard and was curled up dry vomiting. The EM2 radioed back that they were having trouble getting the pump started and wanted us to get them back to the cutter because they were sick. Unfortunately, with the rough weather we couldn’t keep risking sending the small boat out in this weather so they were going to have to tough it out for a while. At around 0400 we were able to send over several other guys to relieve them and to see if they could get it started and the DC1 and EM2 came back to *Dependable.* They were in such bad shape from being sea sick, wet, and cold that they almost fell into the water trying to get back onboard and needed help up the ladder from the small boat. When I reached down and grabbed one of them he was like dead weight.

I think the 4 *Raw Faith* crewmen that came with us will never forget the Thanksgiving feast they had with us!

In the next two pictures you’ll see the enlisted men eating their Thanksgiving Dinner on the crews mess deck after going through the chow line and in the following picture FS1 Rojas setting up the wardroom and getting ready to serve the officer's their Thanksgiving Dinner. I didn't get a picture of the chief's mess but it's better than the enlisted mess but not as good as the officer's wardroom.

There were basically three different groupings of people on *Dependable,* just like on every Coast Guard and Navy vessel. They are enlisted, chiefs, and officers. All three relax, berth and dine in separate areas. And in the case of dining, although there is a huge difference in presentation and eating tableware, the meals are identical and come from the same galley. It's just one of the many differences that you'll find between enlisted, chiefs, and officers aboard a cutter. And since I have experienced all three over my career I just wanted to point it out.

MK1 Resko, FN Sandler, MK1 Ward and EM3 Lesniak enjoying their Thanksgiving dinner

FS1 Rojas setting up the wardroom's Thanksgiving Dinner

Thanksgiving dinner 4 minutes later after taking a roll

After my tour as an officer afloat I would highly recommend taking a commission or becoming a warrant officer. I explain this in detail in a book on leadership I wrote titled "Thoughts on Being A Chief Petty Officer." Basically, it boils down to money. But you also have the perks while active like the wardroom and one or two man berthing to name just two. And if you're going to work why not make more money and live better? But I have strayed from the story and will let you read my other book for more on that.

Setting up a tow is time consuming. The tow line needs to be faked out and made ready. Then the line needs to be passed over to the vessel you are going to tow. This can be done by throwing a messenger line over using a monkey's fist, shooting it over with a special adapter fitted at the end of a rifle, or by taking it over on a small boat. Once that's done you need to haul the line over. This becomes very hard as the heavy tow line, especially as it's in the water, gets faked out more and more. From there it's attached to the vessel being towed and the slack is slowly taken out. This becomes dangerous as it strains to take the weight of the vessel being towed. Once that is completed a 24 hour tow watch is set at the towing bit to make sure the line doesn't part or get enough slack in it that it fouls our propellers.

In the next picture you can see we're trying to get the tow line set up and you can barely make out the *Raw Faith* through the fog. I figured we had about 500' of tow line out on this tow.

Getting the towing hawser ready for a tow

Once we did make it back into Rockland, we turned the tow over to another Coast Guard vessel to take it to the pier. You can see in the following picture the fog had lifted and the seas calmed down.

By now the *Raw Faith* turned into a real news story and we both made the papers. It wasn't so much because of the rescue, but more so because there was a lot of finger pointing as to the sea worthiness of the *Raw Faith* and if it should have even been out there. Their captain was saying his vessel was totally sea worthy and could have ridden out the storm without a problem. The Coast Guard and state officials on the other hand were saying it wasn't sea worthy and had no business being out there.[4]

Raw Faith

The remainder of the patrol was quiet. Some patrols were like that; just sitting around out in the middle of the ocean waiting for something to happen while inspecting fishing boats. We made the best use of the time conducting damage control drills and performing maintenance. On the 1st of December we pulled back into our berth at Cape May to spend Christmas and New Year's at home.

2005

17th of January 2005, we once again left Cape May and set course out into the Atlantic Ocean for another D1 fisheries patrol. But we were having trouble with one of our propellers and needed to get that fixed first. The propellers on modern ships operate on hydraulics that actuate each propeller blade. The system is known as controllable pitch propeller, CPP. When you want to go faster you increase the shaft speed and pitch on the propeller blades. When you want to slow down you decrease the speed and pitch.

We knew we had a problem with the hydraulic system on the #2 CPP blade seal, which holds hydraulic fluid in the system and keeps sea water out. The only way to gain access to the propeller blade seal is to haul us out of the water to affect the repair. So a dry-docking was scheduled and we headed to the Coast Guard Yard at Curtis Bay, Maryland.

We didn't get far. Although we had our own problems, we were diverted to assist the 86' fishing trawler *Provider* the day after we got underway. Her rudder had jammed about 80 miles east of Beach Haven, New Jersey. The seas were rough and at about 14 foot swells. Another Coast Guard cutter, the much smaller 87' *Mako,* was first on scene but couldn't connect a tow line to the *Provider* because the weather was so crappy. We arrived on scene, hooked up the tow, and towed the *Provider* into port at about 2 knots an hour.[2] The following photo is of us towing her. You can see it's pretty rough out and the *Provider* is almost lost in the swells at the end of the tow line.

Towing the *Provider*

After dropping off the *Provider* we continued on to Baltimore. The transit to the Yard was interesting. There are two ways to get there from Cape May. We could go the long way and sail into the Atlantic Ocean and up the Chesapeake Bay, or we could go the short way and sail up the Delaware Bay, transit the Chesapeake and Delaware Canal, and down the Chesapeake Bay. We did the latter. I had transited the Chesapeake and Delaware Canal numerous times on other cutters and it was always enjoyable. It is 14 miles long and only 450 feet wide, lined with houses and small towns which made for a very scenic trip.

Going through the I-695 bridges on Curtis Creek

After arriving at the Coast Guard Shipyard we anchored in the creek until 0600 the following morning and got into position to enter what is normally called a dry-dock. But the Coast Guard Yard had modernized and installed a state of the art lift system that lifted the vessel out of the water instead of the typical way of putting you in a lock and pumping all the water out. This new set up was called syncrolift. We were assisted into the syncrolift bay by tug boats that positioned us over a movable cradle. Next, the movable cradle was lifted out by this syncrolift with us on it. It's pretty impressive.

Entering the syncrolift

You can see in the previous picture the syncrolift cable boxes and the bay we had to slide in to. There is also a 270' cutter already out of the water and on its movable cradle. And in the following picture you can see *Dependable* out and on its own movable cradle.

Once resting on these movable cradles the yard personnel could move us around sort of like pushing around pieces of furniture on rollers.

High and dry on our movable cradle

Working on #2 CPP seal

Once out of the water we commenced repairs on the CPP system. By 2200 we were back in the water and anchored once again in the creek. The next day we headed out and into the Chesapeake Bay. From there it was out into the Atlantic Ocean on patrol where there was a storm and nasty weather awaiting us. But even in bad weather someone has to be out there inspecting fishing vessels and waiting for a search & rescue case.

On the 25th of January we pulled into Newport, Rhode Island for food, fuel, and liberty. It was freezing and there was snow everywhere.

It had been a rough week, and the storm and high seas had really worn us out. When you're constantly straining to maintain your balance, it really tires you out. Then it's hard to sleep because you're getting tossed around in your rack. I would always try and wedge some pillows behind me and bend my knee against the rail that kept you from rolling out of your rack.

I never got physically sick while under way. I definitely didn't feel that hot when it got nasty out. But watches still had to be stood, rounds of the cutter made, and work done. There were guys that got sick and looked like crap and I felt sorry for them. They would walk around with a plastic bag and when they needed to vomit, they let it out and kept on going.

It wasn't much better at the pier in Newport, and we seemed to be rocking just as bad as being out in open water. It was so rough when we were heading out on the 27th one of our crewmembers banged his head against a piece of machinery and gave himself a good concussion. We didn't really know the extent of his injury and decided to have him transported back to a medical facility ashore. So a 47' small boat from Coast Guard Station Point Judith, Rhode Island came out and got him.

47' boat from Station Point Judith, Rhode Island

After completing the transfer of our injured crewman we headed back out to sea passing the decommissioned *USS Forrestal* and *USS Saratogo*. Two famous Navy aircraft carriers that had served their country well and were now just waiting to be scrapped. Or if they were lucky turned into a floating museum. We looked at them in awe as we passed their huge superstructures. Then it was on to boarding and inspecting fishing vessels. In the next couple of pictures you can see our boarding team getting ready to go over to a fishing vessel and boarding a fishing vessel on the 30th of January.

These events played out with the boarding team getting issued their weapons, going up to the flight deck, lock and loading their weapons, getting into the small boat, and then heading over to the fishing vessel. You can see by the way they are dressed that everything is bulky and cumbersome. With all that added weight they have to get from our bobbing small boat that's slamming alongside the fishing vessel, up a rope ladder, and onto the fishing vessel. Next, they have to get their gear aboard, which is in waterproof cases, and contain paperwork and tools to measure net sizes. They do this by hoisting them up with a rope. On board they will inspect the vessels licenses, documentation of the crew, conduct a safety inspection, catch inspection, and a net inspection.

Once they finish their inspection they have to lower down their gear and get back into our bobbing small boat. This is usually done by timing the swell and jumping into the boat. Then it's the transit back to our cutter and lifted back aboard. From there it's clearing and turning in their weapon, debriefing the captain, and then the all-important paperwork of documenting the boarding and processing any violations or warnings that were issued. And like I mentioned earlier this is a collateral duty so when they're done, they head back to their normal job on the cutter.

In the following picture is boarding team bravo left to right; Electronics Technician First Class Weir, Gunners Mate Third Class Spalding, Machinery Technician First Class Hennessey, and Boatswains Mate Third Class Dahl.

Boarding Team Bravo

Boarding a fishing vessel

On the 16th of February it was time to head home and we arrived back to cold and quiet Cape May with as much fanfare as when we had left - none.

We were home for about six weeks, which was the normal inport period. But being inport didn't mean we were off. After pulling in the captain routinely gave everyone, unless they had duty, 72 hours off. Duty was stood by everyone, enlisted and officers alike. For junior enlisted it was a 1 in 4 rotation, which meant you had to stand a 24 hour watch aboard the cutter every fourth day. For senior enlisted and officers it depended on how many people were in the duty rotation, but it averaged about 1 in 7. Monday through Friday were work days from 0800 to 1600. And there was a lot to get done before getting underway again. The machinery needed serviced or overhauled, the boat cleaned and repainted where needed, stores and fuel taken on, new crewmembers trained, charts updated for the next patrol area and so on.

Our homeport of Cape May was not just home to a few cutters and a small boat station, it was also home to the only basic training facility currently in the Coast Guard. So we had to look good because we were the example of what every recruit thought a cutter looked like. And there were always recruit company's getting their class picture taken in front of the cutter and dignitaries wanting a tour. This also meant getting the cutter in full dress with flags from stem to stern like in the following picture with the *Vigorous* on holidays.

Full dress for Presidents Day 21 February

After discontinuing basic training at Alameda, California in 1982 Cape May became the only Coast Guard basic training facility. So, everyone who has joined the Coast Guard since 1982 has come through Cape May, including me. Although when I went through it was a sort of watered down basic training because I was in what was referred to as a "prior service company" since I had already served three years in the Army before joining the Coast Guard.

But even prior service boot camp wasn't a cake walk. Because the Coast Guard is a lifesaving service you have to know how to swim. And preferably well enough to be able to save someone who is drowning. So, the swim test was probably the most difficult event for everyone to pass. It was a make or break event. If you couldn't pass it, even after several tries, you were discharged. It sort of makes sense if you're going to be in the Coast Guard and save lives at sea you should know how to swim.

The test was conducted in a huge indoor swimming pool. You started off by jumping into the pool from a raised platform to simulate abandoning a sinking ship. From there you had to swim the length of the pool and back, which was about 100 meters, and then tread water or stay afloat for about five minutes. Once you passed the swim test we were trained in how to turn our pants into a life preserver by filling the legs with air. And all kinds of techniques for staying afloat – just in case your cutter sank. But we don't want to dwell on that thought.

We set sail once again on the 3rd of April. This time we were headed for Little Creek, Virginia for training and qualification certification known as TACT. This basically involved demonstrating everything we do underway and being judged on it to make sure we were doing it correctly. From getting underway to landing a helicopter, it was all evaluated.

Here is a picture of BM3 John Martorelli firing one of the .50 caliber machine guns during qualification. The target he is shooting at was being pulled by a remote controlled jet ski. Unfortunately, John was killed in a car accident years later.

BM3 John Martorelli firing the starboard .50 cal.

When we finished with TACT we headed straight down to Gitmo for fuel and into our operating area.

When we got down into the Caribbean Sea the captain, like on every cutter I had sailed on, tried to give the crew the day off on Sundays. They called it "holiday routine" or "rope yarn Sundays." If Sunday turned into a work day he tried to squeeze it in at some point during the week. And if the weather was nice and the water warm enough swim call was held. The cutter stopped and the screws secured. An armed shark watch was posted on the bridge and the small boat usually lowered into the water to assist with anyone in distress. This next picture is from the 27th of April and is swim call with guys swinging into the water from the single point davit crane used for lowering and raising the RHI small boat. The water temperature was 81 degrees.

Swim call

26th of April we officially relieved another 210 of their migrant duties so they could out-chop and go home. It was always nice when the cutter relieving you showed up on time and you could get on your way home. The relief consisted of an intelligence brief and the transfer of any migrants they had.

29th of April, we stopped off near a remote and deserted island in the Caicos Islands chain and had a beach party complete with a grill. We did some snorkeling and noodled around the beach picking up huge conk shells. There wasn't anything on the island but there were the remains of some pretty good size ships that had either run aground and were sunk there or washed up in the shallows. It was pretty common for migrants to get stranded on these remote islands, so we kept a look out for any signs of castaways.

It was always nice to be sailing around the Bahamas and we had a French interpreter, because Haitians speak French, and a Bahamian Naval officer, since we were operating in and out of their territorial waters, riding with us. The Bahamian officer even showed our cooks how to make fresh conk salad from the conk shells we had found on the island.

1st of May 2005, while patrolling 10 miles south of West Caicos Island in the Caribbean Sea we rescued 132 Haitian migrants loaded on a 50 foot boat. The seas were choppy and it was after dark when we spotted the Haitian vessel. It wasn't until about 2330 that we started to transfer the Haitians aboard our cutter via our two small boats and housed them on the flight deck until we repatriated them in Port-au Prince, Haiti. The Haitians had been at sea for five days and were dehydrated and hungry. They had no food, water, or sanitary facilities on the dangerously overloaded boat when we intercepted them. It wasn't until about 0145 when I went by the galley and the cooks had some sausage, egg, and cheese sandwiches to snack on. They took such good care of us!

In my experience Haitians always looked worse than any other migrants we stopped. And I think this had a lot to do with the fact that they started their journey less prepared than others, their boats were fuller than most, and more importantly they were at sea longer. Cubans were going 100 miles to Florida, Dominicans just 70 miles to Puerto Rico, but Haitians were trying to make it 700 miles to Florida. Add to this that the Haitians were usually abandon after a day or two by their guides and left to fend for themselves made the journey extremely dangerous for them.

To say it was sad and heartbreaking to see the condition they were in is an understatement. They had left everything for a dream and now we were stopping them. But in reality we were saving them from an almost certain death at sea.

Taking Haitians onboard the RHI

The process of interdicting migrants was an all hands evolution. My job was to film and photograph the whole process. As we came upon a suspect vessel I would start filming and taking pictures. The purpose of this was to document the entire evolution from interception to safely embarking them.

When we come upon a suspect vessel of migrants they are going to be either friendly or hostile. As I had mentioned we have had them throw shoes and knives at us trying to get away and we have had them welcome us as their rescuers. I know I keep saying this, but you have to realize that they have sold everything they own to try and make the transit to the United States and we have just shown up and stopped them. They have usually been at sea for a few days and can also become very excited to get off of their boat, all rush to one side, and capsize. A majority of migrants can't swim, and many have perished due to their boat flipping over when approached by Coast Guard small boats. We were told to never jump in after them because they can pull you down with them. When you have multiple people clinging onto you for their life they are going to take you down with them. If they did go into the water the protocol was to throw life jackets into the water for them to grab onto until we could safely remove them from the water.

Once the suspect vessel has been approached, the occupants are calmed, and life jackets are distributed to all aboard. We then start the process of transferring them from their vessel to ours. And this in itself is very dangerous. Because we have to get them from their boat that's bobbing up and down to our boat which is doing the same thing. Then when we get to the cutter they have to climb up a rope ladder to the main deck. They are usually weak, sick from being at sea, and the climb is tough on them. All children and pregnant women are brought aboard first, then the men. As they are brought aboard we search their personal belongings, they are frisked for weapons, given a blanket, flip flops, and simple toiletries.

Migrants being brought aboard

Haitian migrant being frisked

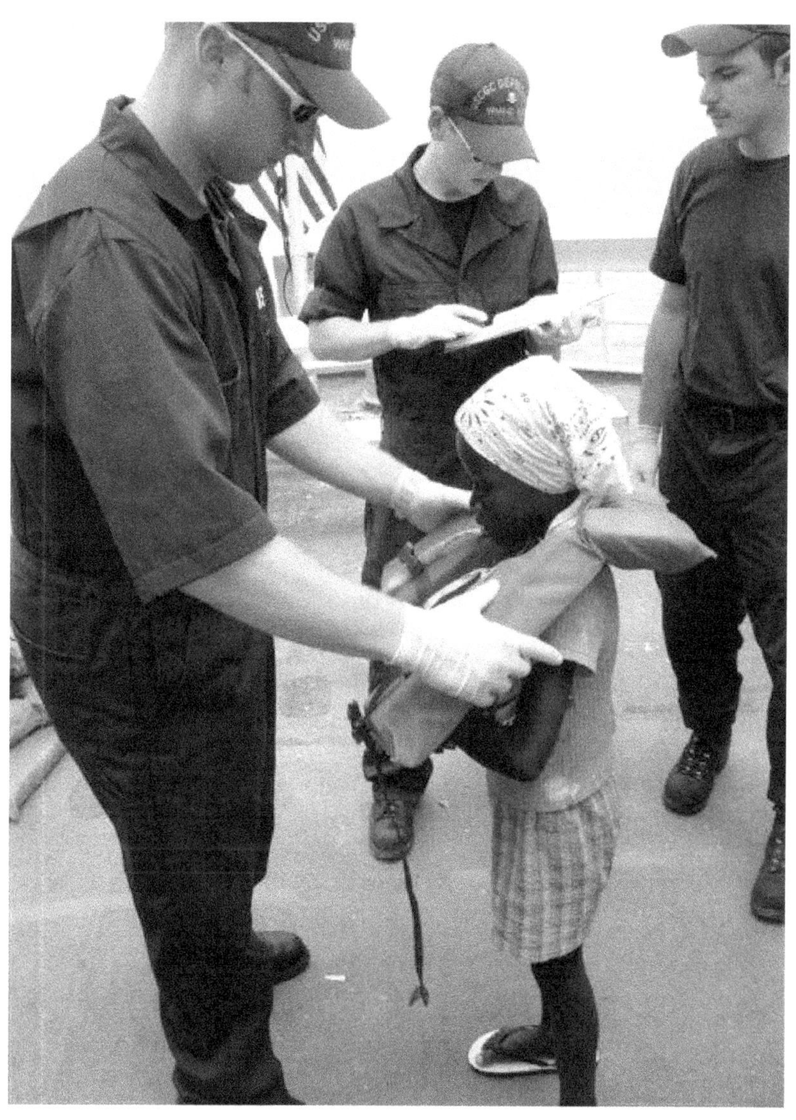

A Haitian child being documented

The flip flops help protect their feet from the hot steel deck because they are almost always barefoot, especially the

Haitians. Medical personnel inspect them for any health issues, they are documented with a wrist ban, and then they are led to the flight deck where we can contain them.

Once settled onto the flight deck we have to stand watch over them. Partially to keep them contained but also to make sure they didn't hurt each other. When it was time for them to leave and they were debarked we had a huge mess to clean up on the flight deck. All their trash had to be picked up, blankets folded and stored, the flight deck washed down, the portable toilets and sinks sanitized, broken down, and stored. It was consuming. If you weren't on a bridge or engine room watch you were tending to migrants. And when you could you took a nap.

Just a note about the little girl in the previous picture. She was picked up with her mother and they were safely repatriated back in Port-au-Prince, Haiti. That picture is of her being fitted with a life jacket for the small boat ride back to Haiti.

4th of May, we repatriated the 132 migrants that we caught on the 1st by ferrying them to shore via our small boats.

9th of May, intercepted a yola containing 100 Haitians in the late afternoon. By 1840 we were about halfway through bringing them aboard. Another long day.

10th of May, A Coast Guard MH-60 helicopter spotted a yola of Haitians and reported about 20 people in the boat. It hovered near the vessel until we could get on scene. Turns out what looked like 20 people actually tuned into 112 Haitians. That's how crowded they were. I think the perception for most of these migrants is that they are just piling into the boat for a few hour journey to America. Not realizing they are going to be out for days, at the least.

This picture is of the yola that the helicopter thought had 20 people and it ended up having 112 on board. Look at it closely and you really can't count more than 30 or so. That's how packed in they are. They are literally stacked on top of each other.

Yola carrying 112 Haitian migrants

12th of May, while patrolling 50 miles northwest of Matthew Town, Great Inagua Bahamas we rescued 116 Haitian migrants in another dangerously overloaded 50 foot boat. After removing the Haitians from the boat we destroyed it at sea. The 116 Haitians consisting of men, women, and children were taken to Port-au-Prince, Haiti and repatriated via our small boats later that day. This group included the little girl getting fitted for a life jacket in a previous picture.

Like with the Dominicans it was always an orderly affair. I was told that the Red Cross gave them some money, personal items, and they were released back into the public.

From there we set a course for Key West to refuel and take on food stores. On the 18th of May we departed Key West and commenced our eight day transit home. On the 26th of May we entered the inlet into Cape May and made berth.

That was my final patrol on *Dependable*, 53 days away from homeport. After liberty was granted, I walked off *Dependable* for the last time. There was nothing unusual about the event, it was like every other time I had departed the *Dependable*. I walked back to the fantail where the quarterdeck was located, slid the lever by my name from green to red indicating I was ashore, saluted the quarterdeck watch stander, walked to the edge of the brow, saluted the ensign, and walked ashore headed to my new assignment.

I wasn't sure if this would be my last cutter or not. I was an officer now with a lot of potential assignments ahead of me, on land and at sea. At this point well over half of my time in the Coast Guard had been spent at sea. But in the end, this would be the last time I crossed the brow of a Coast Guard cutter. I would not go back to sea again. It was a fitting end and I'm glad it was on such a great cutter with such a great crew.

Walking off *Dependable* ended my time at sea; leaving me with my permanent silver cutterman's pin as an enlisted man, gold cutterman's pin as an officer, sea service ribbon with three bronze stars, and a whole lot of memories!

References

1. Asbury Park Press, 14 Jan 2004, page 20

2. Asbury Park Press, 20 Jan 2005, page 3

3. South Florida Sun Sentinel, 14 Sept. 2003, page 3

4. The Burlington Free Press, 27 Nov. 2004, page 18

About the Author

Ed Semler retired from the United States Coast Guard in December of 2007 with over 25 years of military service in both the United States Army and United States Coast Guard. In the United States Army he was an enlisted man and was honorably discharged as a Specialist Four (E-4). While in the United States Coast Guard he was enlisted, obtaining the rank of Master Chief Petty Officer (E-9), was commissioned as an officer, and retired as a Lieutenant (O-3E).

After his military career Ed dabbled in teaching at a Vocational Technical School and was a self-employed plumber for several years. As a past time he enjoys writing and playing the guitar, bass, piano, and harmonica.

Fully retired he resides in Schulenburg, Texas with his wife Jana, a retired Air Force senior master sergeant. Please feel free to check out Ed's other books at www.edsemler.com or email him at mkcm378@gmail.com

His other publications are;

"Around The World," a memoir of his 25 years of service as an officer and enlisted man in the U.S. Army and U.S. Coast Guard

"U.S. Coast Guard Cutter Sherman (WHEC-720) Circumnavigation Deployment 2001" which details the *Sherman's* historic circumnavigation of the globe and deployment to the Persian Gulf in 2001

"The Three Gunsallus Brothers" a story about fighting for Pennsylvania during the Civil War

"Sam Houston & Napoleon Bonaparte Meet On The Civil War Battlefield" a true story of the Walker brothers

"Thoughts On Being A Chief Petty Officer" a take on military leadership

"Fighting For Pennsylvania In The Early Years 1763 to 1783 – The Story of Captain Thomas Askey And Lieutenant Richard Gunsalus Of Cumberland County"

"Joe Semler Playing Baseball in the 1920's &30's"

"Alice Springs Australia Adventures In The 80's"

www.ingramcontent.com/pod-product-compliance
Lightning Source LLC
LaVergne TN
LVHW010625100826
845148LV00014B/3119

* 9 7 8 1 7 3 7 6 4 7 2 0 1 *